Renan Campos

Technological modernisation of low-income housing

Renan Campos

Technological modernisation of low-income housing

A look at NBR 15.575

Imprint

Any brand names and product names mentioned in this book are subject to trademark, brand or patent protection and are trademarks or registered trademarks of their respective holders. The use of brand names, product names, common names, trade names, product descriptions etc. even without a particular marking in this work is in no way to be construed to mean that such names may be regarded as unrestricted in respect of trademark and brand protection legislation and could thus be used by anyone.

Cover image: www.ingimage.com

This book is a translation from the original published under ISBN 978-613-9-72883-1.

Publisher:
Sciencia Scripts
is a trademark of
Dodo Books Indian Ocean Ltd. and OmniScriptum S.R.L publishing group

120 High Road, East Finchley, London, N2 9ED, United Kingdom
Str. Armeneasca 28/1, office 1, Chisinau MD-2012, Republic of Moldova, Europe
Printed at: see last page
ISBN: 978-620-7-91417-3

Renan Salvate Campos

Graduating in civil engineering from Puc-Rio in 2013. Outstanding academic student in the 1st term of 2008. Monitor of programming, introduction to engineering and topography subjects. Internships in the areas of property development, cost control and construction planning and management.

Ficha Catalográfica Campos, Renan Salvate

Technological modernisation and improving the quality of low-income housing / Renan Salvate Campos; Supervisor: Carlos Éden Mesquita - Rio de Janeiro: PUC, civil engineering department, 2013.

v., 71 f.: ill.; 0,73 cm

1.   Final project for course completion (undergraduate) -
Pontifical Catholic University of Rio de Janeiro, Department of Civil Engineering

Includes bibliographical references

1. civil engineering - final design 2. performance standards 3. affordable housing. 4. Civil construction. I. Mesquita, Carlos Eden. II Pontifical Catholic University of Rio de Janeiro - Civil Engineering Department. III. Technological modernisation and improving the quality of low-income housing.

To my parents, José Renato and Fátima Suélia, for their unconditional love and for teaching me to always dream and persevere.

Thank you

2

I would like to thank my supervisor Prof. Carlos Éden S. Mesquita for his trust, encouragement and knowledge in helping me to carry out this work.

To PUC-Rio and all its teaching staff for the technical and human teachings that are essential for the formation of a good professional.

To my friends Antonio Rossano, Lívia de Castro, Mayara Lobo, Nathália Machado and Rustam Mesquita for the tears and smiles shared throughout my journey.

To my parents for their trust, education and affection at all times.

To all my friends and family who have contributed to my academic and personal development over the years at university.

Summary

Campos, Renan Salvate; Mesquita, Carlos Eden. Technological modernisation and improving the quality of low-income housing. Rio de Janeiro, 2013. 71p. Final graduation project - Civil Engineering Department, Pontifical Catholic University of Rio de Janeiro.

The Brazilian reality is represented by a housing deficit and an increase in the population of cities. For this reason, a number of low-cost housing projects have emerged over the last few years, but those responsible for their design and construction have favoured profit over user comfort and safety. In order to establish criteria for improving the quality of homes, the NBR 15575 technical standard for housing performance was drawn up, allowing new technologies to be applied to construction with the aim of improving the quality of life of residents. This project interprets this new standard and presents possible solutions for meeting its requirements.

Key words

Civil Construction; Performance; Performance Standard; Quality; Affordable Housing; Technological Modernisation

Summary

CHAPTER 1

Introduction

The performance standard for residential construction projects came into force in July 2013. Codenamed NBR15575 Edificações Habitacionais - Desempenho, it represents a major advance in the country's construction industry, allowing for better quality projects and guaranteeing end consumer satisfaction.

For users, the standard makes it possible to know whether the property they are buying has the comfort, stability, structural safety, fire safety and useful life they want. In addition to being able to choose your property by performance rating from a range of options.

For developers, designers and builders, it's a challenge. With society and the economy evolving, new technologies and the search for ever lower costs, it has become necessary to promote new conditions for making projects viable, investments in machinery, better construction processes and labour skills to guarantee the sustainability of the construction industry.

To this end, the performance standard establishes various comparative, objective and quantitative parameters. In this way, it seeks to interrelate all the links in the economic chain involved in the project, reducing uncertainty, reducing predatory competition and differentiating companies in the market. Evaluating construction systems promotes the evolution of all the components in the chain, as well as making it clear to the consumer which companies are thinking along the same lines.

Unlike the other standards, NBR 15575 provides for the concept of the behaviour in use of the components installed in the building, thus ensuring that users' requirements are met not only during the construction period, but throughout the building's useful life. For this to happen, project management is required, involving all those responsible: developers, designers, builders, suppliers, the building itself and users. This integration, when carried out partly during the planning phase and then throughout the development of the building, can be achieved by

he development of the building promotes an improvement in quality and the optimisation of costs. In addition, while the other standards set products to guarantee the characteristics of the building, the performance standard promotes the development and application of products that meet the needs of the building, and consequently stimulates the emergence of new technologies, materials and construction methods.

The expectations of the market and the technical community with the application of this standard go beyond improving the quality of construction itself, it represents a new milestone in which good suppliers will consolidate their hold on the market and others will have to look for alternatives. It is hoped that the consumer will gain in every respect, the quality of new projects could be higher than the current model, illegality will tend to decrease as it will no longer be attractive and the market can regulate itself at a high level.

Based on international standards, the standard was built according to user needs and exposure conditions:

1. General requirements

2. Requirements for structural   systems

3. Requirements for flooring systems

4. Requirements for internal and   external vertical sealing systems

5. Requirements for roofing systems

6. Requirements for installation systems

Each part has been organised according to safety, habitability and sustainability requirements:

• Safety: mechanical performance, fire safety, safety in use and operation

• Habitability: Watertightness, thermal, acoustic and lighting performance, health, hygiene, air quality, functionality, accessibility and tactile comfort.

•   Sustainability: Durability, maintainability, environmental suitability

This standard applies to any residential building with any number of floors, with the exception of buildings with up to five floors. The only buildings excluded from the standard are those built prior to its publication, existing buildings, renovation or retrofit work and temporary buildings.

CHAPTER 2

General performance requirements

2.1.1 mplementing the work

When drawing up the project, attention must be paid to obtaining the land and setting up the construction site itself, mainly because the site chosen is part of the success of the project. For this reason, the standard defines a series of requirements regarding the assessment of geomorphological aspects when choosing the site and implementing the work: Risks of mass movements and flooding, the presence of collapsible, expansible soils or craters in deep layers, the possibility of soil deconfinement and the existence of soft clay in deep layers, reactions caused by lowering water tables, the effects of pile groups and earthworks vibrations, checking for the presence of gases in the air or confined underground due to disused landfills or dumps and other neighbourhood effects that could cause physical damage or inconvenience to the users of the dwellings.

To this end, it is worth consulting the town hall, the fire brigade, civil defence and the local population in order to verify the possibility of the effects described above occurring. When any occurrence is identified, it is recommended that the project provides for a cost-effective treatment of the problem, minimising the need for the end user to interfere in the solution.

It is worth remembering that the entire interference study and the solutions found, as well as environmental compensation, must be described in the use, operation and maintenance manual.

2.2.5 health, hygiene and air quality

2.2.6 Health and sanitary code

Housing must comply with the local health code and ANVISA regulations. Amongst these, it is also necessary to design in such a way as to avoid the presence of insects, rodents, micro-organisms and other vectors of disease, whether contagious or not. In addition, the project can be designed to prevent suspended particulates, vehicle fumes and especially toxic gases from entering the building.

Some strategies for complying with the standard are: Build with materials that do not retain moisture, preventing the proliferation of fungi and micro-organisms; design doors, windows, rooflights and openings that favour ventilation and sunlight; design ventilation systems, preferably natural, for garages; ensure that roofs, façades and windows are watertight; avoid environments and structures that encourage the accumulation of water; rubbish bins with watertight, washable, ventilated and locked linings; avoid niches or architectural cut-outs that could serve as supports for bird nests or insect refuges; in places without a sewage system, the project can provide for mini-treatment plants according to your needs.

All water and sewage installations must comply with their respective technical standards and all design information and instructions for maintaining and cleaning the systems must be contained in the use, operation and maintenance manual.

2.2.2. Pollutant content

All internal gas-burning equipment, such as the water heaters commonly used in flats today, must not have concentrations of more than 0.5 per cent CO_2 and 30 ppm CO.The Brazilian Electrical and Electronic Industry Association carried out a comparative study in 2007 of the water heating systems available in Brazil and came to the conclusion that electric showers are more economical both in installation and in use, presenting the results according to each piece of equipment per shower in a home:[i]

Heating system	Installation costs	Cost of equipment	Cost of use
Electrical	R$40,00	R$25,00	R$33,61
Natural Gas	R$900,00	R$350,00	R$79,69
Solar	R$3495,00	R$1140,00	R$52,76

Table 1: Comparison of heating systems

It's up to the developer and designers to check which system is best for their project, remembering that all the relevant information on installations and maintenance guidelines must be contained in the use, operation and maintenance manual.

2.2.3. Gas and insect tightness

In addition to complying with NBR8160, all sewage installations must be designed to avoid loss of water seal and resiphonation, thus preventing the return of water, waste or odours from the installations.

2.2.4. Risk of drinking water contamination

Hydraulic systems must be completely isolated from each other and especially from the sewage system, thus avoiding the risk of contamination. The standard calls for physical separation between them, the materials used to drain the water must not transmit toxic substances or heavy metals, parts of apparent installations must be washable and impermeable, preventing the formation of colonies of bacteria and other micro-organisms, and equipment must not allow water to puddle inside.

Iron and copper pipes suffer greatly from the weather, contaminating the water and potentially harming the health of users, and it is preferable to use PVC or stainless steel pipes for drinking water.

Special attention should also be paid to water tanks, designed without direct contact with the ground and with airtight lids, which should always be clean and properly maintained. Ideally, maintenance should be carried out annually and preferably in the winter months, as consumption is lower[ii] . É

It is important to hire a specialised company for the job and also to carry out water potability tests in specialised laboratories.

The operation and maintenance manual must contain guidance on the maintenance of equipment, installations and tanks, as well as how often it should be carried out.

2.3 Environmental suitability

As common sense and new environmental legislation dictate, buildings should be constructed in such a way as to have as little impact on the environment as possible. Even though NBR15575 does not specify strict terms for environmental preservation, it is worth paying attention to a few points such as:

Construction sites can avoid land close to watercourses, respect marginal protection strips and carry out environmental compensation whenever requested.

Developments built from locally extracted inputs or that will operate from local sources can provide for sustainable management of these resources.

Avoid using wood, but if necessary, use legal species from plantations authorised by environmental agencies.

Carry out an efficient and rational waste management plan on the construction site in accordance with CONAMA 307 and 448: The former regulates the classification of waste and its destination, as well as guidelines on reuse, storage, transport and disposal at legal sites.[iii] The second defines the characteristics of landfills, overflow areas and waste management plans and states that they must be implemented.[iv]

At the design stage, it is important to check with manufacturers about the life cycle of their materials, opting for those that can be recycled and are produced with respect for the environment. In addition, design the building in such a way as to maximise the use of natural lighting and ventilation, thus saving electricity over the life of the project.

2.3.1. Sustainable locations

The location of the project has a number of influences on its development and running costs. Both during its execution and utilisation, the site needs to meet the requirements of water, energy and the logistics of raw materials and human beings. These resources can be taken from the site or transported to it, always seeking a balance between the three environmental, economic and social premises. You can also take into account how much of the land will be developed and how much will be preserved, an area that can be calculated right from the design stage. It is worth assessing how much of the natural heat will be used both in open areas and on roofs, seeking to minimise artificial heating or cooling.[v]

2.3.2 Choice of terrain

The land can be chosen in such a way as to avoid sprawling buildings in suburban areas that are still unspoilt by nature. The main aim is to increase density, i.e. to house more people in less space, so you need to grow upwards (or downwards) rather than sideways. The reason for preserving open and wooded spaces is to maintain a natural refuge for generations to come.

2.3.3   Rationalisation of water use

According to the specifications of standard NBR15097, it is recommended to use water-saving equipment, such as toilets with a reduced or varied volume, self-closing taps and reliable, well-regulated tank floats. It is worth noting that restricting the volume of water available too much can have the opposite effect, as a flush with too little water may not be enough to drain the waste, forcing the user to flush several times.

SABESB[vi] has drawn up a table of energy-saving equipment and presented some alternatives, which can be seen below:

Conventional equipment	Consumption	Economising equipment	Consumption	Economy
Bowl	12 1/flow	vdr basin	6 1/flow	50%
Shower	0,34 1/s	Flow restrictor	0,13 1/s	62%
Sink tap	0,23 1/s	Air	0,10 1/s	57%
Tank tap	0,28 1/s	Flow regulator	0,13 1/s	50%
Urinal	2 l/s	Automatic valve	1 l/s	50%

Table 2: Comparison of plumbing equipment

2.3.4   Risk of soil and groundwater contamination

It is recommended that during design work the occurrence of settlement of embankments and foundations is checked in order to foresee damage to existing sewage systems or those that are to be built.

will be built in the future, thus avoiding soil contamination. This risk can also be associated with the construction stage, where the waste management of both the construction itself and the workers who work on it must be disposed of in accordance with environmental legislation, minimising any type of soil contamination, even if superficial.

2.3.5.   Water utilisation and reuse

The standard suggests that all water from sanitary facilities should end up in the sewage system. If this does not exist, it can be connected to a coherent treatment system to avoid contaminating the soil, as already mentioned in the "Health and sanitation code" section above.

It is recommended that grey or rainwater be reused to save potable water in areas where there is no need to use it, for which the standard establishes the following parameters:

- Absence of total coliforms in 100 ml
- No thermotolerant coliforms in 100 ml
- 0.5 to 3.0 mg/L of free residual chlorine when this component is used to disinfect water
- Turbidity less than 0.2 uT

- Apparent colour less than 15 uH
- pH adjustment to a range between 6 and 8

Before opting for a reuse system, it is very important to carry out a detailed study of the return on investment. This study should take into account the amount of water generated by users, the local rainfall throughout the year and the building's collection capacity, the amount of water required, the size of the reservoirs and the type of water used, choosing which equipment it will be used in, thus defining the need for treatment to meet the recommendations of the standard. It is worth remembering that the system will need periodic maintenance, so the project should opt for systems that are easy to access, easy to maintain and write down all the relevant information about the system in the use, operation and maintenance manual.

CHAPTER 3

Structural performance

In general, the other technical design standards consider it important that the building withstands the stresses imposed by nature and remains standing with its characteristics unchanged. The performance standard also takes into account the effect of user use on the structure. In addition to evaluating the effects of the Ultimate Limit States, the performance standard also considers the Utilisation Limit State, since this limit can damage the other systems installed in the building, compromising user safety.

3.1. Stability and resistance of the structural system

In addition to complying with Brazilian design standards, the performance standard also regulates certain tests that can be carried out to analyse the resistance of a given material, which has not yet been standardised, thus allowing it to be used if it is more suitable for the project. In addition, when dimensioning pillars, walls, beams and other structural elements, accidental loads acting in conjunction with dynamic actions and the installation of suspended loads on these structures must be taken into account, i.e. the installation of cupboards, shelves and racks on walls and beams, roofs, the possibility of future installation of solar heaters, swimming pools, among other overloads. Remember that all considerations, information on where these installations may or may not be located and the load capacity must be included in the use, operation and maintenance manual.

3.2. Displacements and cracking states of the structural system

In addition to complying with the design standards, the performance standard also requires compliance with certain limitations on the deformation of the structure, with a view to user comfort. Walls and beams cannot have a displacement greater than their span size divided by 250, as this causes psychological insecurity and visual discomfort. It also provides for displacement limits to avoid malfunctioning frames, doors, windows, detachment of ceilings and seals. It is up to the designer to size his parts to withstand these displacements throughout the project's useful life, also providing for the actions already mentioned in the previous topic. It is up to the executor to guarantee the quality of the service by complying with the design specifications.

3.3. Resistance to soft and hard body impacts

In order to prevent damage from vandalism and attempted break-ins, the structure must be prepared to withstand these impacts without being pierced, collapsing, becoming unstable, cracking or suffering any other failure that jeopardises its usability. Impact energy is measured in Joules and has different values depending on where the impact occurs, internally or externally, and the type of sealing, structural or otherwise.

In general, steel frame walls, dry wall or other more fragile systems must undergo the tests stipulated in the standard before being applied. Concrete, steel and some masonry structures have no problems with this type of impact. Balcony railings must follow the specific regulations of NBR 14.718. Testing is recommended for all other materials.

The standard also regulates the impact of a soft body on exposed pipework, in which case the specified test must be carried out as it depends not only on the material but also on the assembly of the system.

3.4. Support capacity for suspended parts

All the walls, ceilings and ceilings in a building must be able to withstand the fastenings of equipment that may be hung from them in the future, such as pictures, mirrors, nets, pipes, shelves, fans and so on. The standard specifies allowable displacement values for each performance level and load increase factors to be considered. These values minimise or prevent the appearance of cracks or rupture of the fixing element.

It is important for the designer to advise the user in the use, operation and maintenance manual on the best fixing system to use. Drywall fixing systems are very different from ordinary masonry fixing systems and are very little known in Brazilian culture. If non-conventional walls are chosen, it is important to provide guidance on how to use them correctly to avoid damage that could jeopardise the useful life of the element.

CHAPTER 4

Fire safety

Fire safety consists of three stages: prevention, escape and combat. The standard takes these three items into account when drawing up guidelines for personal and material safety. Consideration is given to ways of hindering the start of a fire and its spread, analysing the fire resistance of the elements, the possibilities of escape routes, the availability of initial fire-fighting equipment and the ease of access to the fire brigade if necessary.

4.1. The start and spread of the fire

All electrical, gas and lightning protection installations have specific regulations on preventing the ignition of fires. It is up to the performance standard to provide guidance on how difficult it is for flames to spread.

Neighbouring buildings must have a safety distance in accordance with local legislation, and the building may have area isolation systems consisting of fire doors.

Fire spreads horizontally and vertically. In the case of horizontal propagation, it is important to check the spacing between adjacent buildings with parallel windows and doors, as these are the places where fire tends to escape most intensely. It is also interesting to design forms of insulation between semi-detached buildings, preventing fire from passing through walls, floors, ceilings or roofs. In the case of vertical propagation, the fire tends to seek out areas with large openings, such as staircases, shafts, lift shafts and floors with mezzanines, so these areas can be designed with anti-flame and anti-smoke insulation.

During the spread of a fire, even with the measures taken, it is necessary to evacuate the premises. To this end, the escape routes must comply with their respective standards and be well lit, labelled, with free and accessible access for any user.

4.2. Structural performance

In situations where the fire has reached large proportions and high temperatures, there is concern about the possibility of damage to the structure. For this reason, the materials used in the structure must have a Required Fire Resistance Time as specified in NBR 14.432, and if they do not, measures must be taken to meet these parameters.

Structural elements must be watertight, preventing the release of gases through cracks, thermally insulated, ensuring that the propagation of heat by conduction and radiation to the opposite face does not reach $180°C + $ Ambient, and stable, maintaining their physical integrity without compromising their functionality. If it meets the first criterion, it is considered fire stable, if it meets the first and second, flameproof, and fireproof if it meets all three criteria.

In general, buildings must withstand 30 minutes of fire as specified above for both horizontal and vertical propagation. Only multi-family buildings over 12 metres high must withstand 60 minutes, 23 metres 90 minutes, 30 metres 120 minutes, 120 metres 180 minutes and basements equivalent to the time of upper floors, with at least 60 minutes for up to 10 metres deep and 90 minutes for greater depths.

4.3. Reaction to fire

The standard presents a fire resistance classification that can be followed when choosing the materials to be used in the building. It also presents the possibility of a floor system to be installed in order to prevent the vertical spread of flames, consisting of a waterproofing layer, thermal and acoustic insulation and a subfloor layer.

There are elastomeric sealants[vii] which, if applied to the interface between slabs and walls, or between pipework and slabs, can contain flames for up to 4 hours. Other compounds can be prepared as a mortar and make a protective layer on the floor, guaranteeing the necessary protection. Metal structures can also be given sacrificial tinted paints to withstand the standardised time.

Regarding plumbing, the standard also states that for diameters greater than 40 mm, the hole left if the pipe is consumed by fire must be sealed automatically by means of a sealing system.

Natural or mechanical ventilation pipes that cross slabs must also be fire-resistant by means of fire dampers with automatic activation on each floor. They can be activated by smoke detectors in accordance with NBR 17.240 or by heat sensitivity. There are systems[viii] for small pipes, such as automatic valves that block the pipe, for tunnels in which metal sheets melt at temperatures that could present a risk, entumecent grilles that block the passage of air for a set time, guaranteeing the time needed for evacuation, metal grilles with fire resistance and automated closure, among other intervention equipment. When it is not possible to install registers, the pipework must have a fire resistance of 120 minutes, not less than the building's time. All sealing systems must be approved by NBR 6479.

Lifts and staircases are favoured paths for fire, as they have large vertical clearances, and as a point of union between all the floors of the building the entire interface between the slabs must be prepared to prevent the spread of flames by means of fire doors that withstand a time equivalent to the building's endurance.

4.4. Firefighting

As a rule, buildings must have a reserve of cold water that cannot be used for user consumption and must be sufficient for initial firefighting in accordance with local legislation. In multi-family buildings, the fire reserve must be 20% of the consumption calculated for use considering 2 days of storage in accordance with

NBR 5626, and the hydrant system must be interconnected to the public hydrant if available.

In addition, according to standard NBR 12693 fire extinguishers must be positioned, identified and ready for use.

4.5. Relevant information on fire prevention

Although the building is prepared to prevent and fight fires, it is essential to avoid starting a fire in the first place. It is important that the use, operation and maintenance manual includes information on the improper use of plugs to connect electronics, the accumulation of flammable products inside the dwellings, care when lighting fires inside the units, care when disposing of cigarette butts, guidance on avoiding the obstruction of fire-fighting equipment and escape routes, information on the importance of always checking the validity of fire extinguishers and carrying out regular maintenance, information on the protection systems installed during construction and the need to replace sealants or parts throughout the life of the building.

CHAPTER 5

Safety in use and operation

Accidents in the home happen all too often, but they only occur because there are risks associated with the activities the user was carrying out. The standard establishes requirements that minimise these risks in order to prevent the most common domestic accidents, guaranteeing safety and comfort for residents.

Most risks are reduced mainly in the design phase, where the materials used in each situation are defined, while the execution phase is responsible for ensuring the quality of the finishes, avoiding sharp corners and edges, exposed wiring, unprotected pipework and other risky situations.

5.1. Building system security

The building itself must have a safe structure, without the possibility of ruptures, instabilities, vibrations or displacements that threaten the resident, and must not have sharp or perforating parts during use.

The project can prescribe structures and materials to prevent falls from eaves, balconies and roofs, access control to the most critical and accident-prone areas, regularised floors, ramps and stairs, quality window frames, electrical and plumbing installations to prevent damage during handling.

To this end, it is important that the use, operation and maintenance manual contains information on the periodic maintenance of structural systems and extra care in high-traffic communal areas, on the risks of hanging clothes and potted plants on windows and balconies, as well as overloads on walls and roofs in disagreement with the information presented in "Stability and resistance of the structural system".

5.2. Safe use of flooring

Slips are accidents that can have serious consequences, especially for the elderly, which is why the standard regulates that floor finishes must have a coefficient of friction capable of preventing the user from losing stability even when wet.

Floors in wet areas can be slip-resistant in accordance with standard NBR 13818. Some very smooth floors such as glazed or enamelled ceramics are very slippery and do not meet the standard, which is why it is recommended to use a complementary system to increase the coefficient of friction. The Anti-Slip System®[ix] uses a chemical that modifies the molecular structure of mineral floors, making the surface non-slip in the presence of water, soap and oil, does not alter the colour or durability of the tile and complies with the standard.

Although there is a system that increases the floor's coefficient of friction, it may be more economical to specify a floor that is already naturally non-stick, this decision is up to the designer.

The standard also recommends that rounded edges and, if possible, rubber flooring should be used in stairwells, ramps, swimming pools and playgrounds. It also suggests greater flatness, with gentle drops towards drainage systems, preventing puddles without causing tactile and visual discomfort.

Staircases must comply with fire regulations, such as mirrors of the same height and tread of the same width, and preferably a landing every 14 steps, always with handrails on both sides.

5.3. Safety when using the roof

As well as not allowing roof tiles to move and water to permeate roofs, it is important to provide structures for the maintenance of the equipment that is there. The standard recommends designing supports to hold lifelines where workers will fasten their safety harnesses, designing closing beams or another specific structure to support the load of scaffolding that may be hung to carry out maintenance on the façade, grounding the metal structures and equipment, isolating the lightning protection system, making a path between the structures for the maintenance of gutters, tiles and ventilation pipes, among other things.

All information regarding maintenance and use of the roof as a support point for maintenance of other areas must be specified in the use, maintenance and operation manual. If there are solar or gas central heating systems, or other equipment that requires specialised maintenance, the manual must also contain the necessary information for hiring technical advice.

CHAPTER 6

Functionality and accessibility

At this stage of the analysis, it is important to emphasise the standard's concern for the user's freedom of movement and comfort. Functionality and accessibility requirements include regulations on minimum space for furniture and the provision of extensions to certain areas.

The entire building must be designed, in terms of its volume, taking into account the appropriate spaces for carrying out the activities inherent to the site.

6.1. Minimum headroom

Considering that the ceiling height is the distance between the finished surface of the floor and the finished surface of the ceiling of a floor, the minimum height allowed is now 2.50 metres, allowing a 20 cm reduction for areas such as corridors, toilets, changing rooms and pantries.

Even in situations where there are exposed ceiling beams or other architectural peculiarities, 80 per cent of the ceiling height must be 2.50 metres, and at no time less than 2.30 metres.

This request was prompted by the emergence of projects with very small dimensions, aimed at taking advantage of the vertical space allowed by legislation and gaining more units on the upper floors by saving centimetres on the lower floors, but as soon as financial interest overtook common sense, the ceilings became low enough to cause discomfort for users, sometimes even preventing the installation of ceiling fans or light fittings.

6.2. Minimum availability of spaces for use and operation

The use of environments is directly linked to the arrangement of furniture and the way in which the user interacts with it. The standard has drawn up table F1 (NBR15575 Part I) specifying which equipment and furniture are standards of

This means that the project must be carried out with awareness of the existence of this furniture in the rooms, thus guaranteeing adequate space for its use.

Among them are

- Double bedroom - 1 double bed, 1 wardrobe, 1 bedside table
- Living room - 1 2 or 3-seater sofa, 1 cupboard or bookcase, 1 armchair
- Kitchen - 1 cooker, 1 fridge, 1 sink, cupboard over sink, cabinet, 2-seater dining table
- Bathroom - 1 washbasin, 1 shower, 1 basin

Table F2 also establishes the standard dimensions of each piece of furniture and equipment and the space allocated for circulation between them and between walls.

It is worth emphasising the importance of indicating in the use, operation and maintenance manual the standards chosen during the design and assembly options for the internal environment of the home in order to guarantee the performance expected by the user.

6.3. Operation of water and sanitary installations

All plumbing installations for direct user use must be regulated to optimum conditions without discomfort caused by high pressures or reduced flow rates in taps, basins and showers. They must also allow access for maintenance tools, especially in siphons over sinks, which are usually hidden inside cupboards. They must be able to receive couplings for other equipment such as garden hoses and water filters, as well as maintaining standardised gauges and threads, allowing the user to carry out maintenance with products easily found on the market.

As stated in the "Health, hygiene and air quality" chapter, at no time should the user come into contact with wastewater and the sewage treatment system, except for maintenance in accordance with the appropriate instructions, and for this purpose the sewage appliances must function properly throughout the project's useful life. Remember that the maintenance procedure must be specified in the use and operation manual.

The rainwater collection system, whether for reuse or disposal, must be correctly sized to prevent water from accumulating on roofs. Water collection systems using vortices in the drain help prevent blockages from leaves and other materials carried by the wind.

6.4. Suitability for users with physical disabilities or reduced mobility

It is recommended that the housing project contain a minimum number of units ready to receive special users, in addition to all common areas complying with NBR 9050. The aim is to guarantee mobility for everyone within the development.

It is worth noting that for projects that are part of the Minha Casa Minha Vida programme, there is Law No. 12,424 of 2011, which regulates the need to reserve at least 3% of the housing units in multi-family developments for special users. Bill No. 650 of 2011 is currently before the Senate, which seeks to make it compulsory for the builder to adapt more units if there is a demand from people with special needs during the sale of the units.[x]

6.5. Expansion of development housing units

For those projects that present special conditions on common floors that can be expanded, the construction company or developer can provide the user with sufficient information on how to carry out this expansion and what factors need to be considered. It is important to provide information on all the executive projects and as-built projects, such as architectural, structural and installation projects, so that new development projects can be drawn up, recommendations on the use of available regional resources, construction methods and technologies used and available, as well as other information relevant to the type of project. This guidance should be sufficient to at least guarantee the performance of the existing building, while meeting the same levels of functionality.

CHAPTER 7

Tactile and anthropodynamic comfort

Ergonomics must be present in housing projects. By studying the dimensions of human beings and their movements, it is possible to make the right choice of metals and window frames, promoting better use for adults and guaranteeing safety for children. It is also important to study the sensations caused by floor vibrations, the slope of ramps, the speed of lifts and the dimensions of stairs. For people with special needs, the correct installation of support bars, tactile flooring and the provision of adequate manoeuvring areas.

7.1. Floor flatness

Except for areas that have been designed this way for architectural reasons, finished floor areas or surfaces that will receive floors must have a flatness of more than 3 mm per 2 metres. These irregularities can cause discomfort when walking, visual damage and inadequate water run-off.

7.2. Manoeuvring devices

Elements for manoeuvring equipment must be consistent with the human anatomy; registers, latches, locks and valves must be easy for adults to use and not present a risk of cuts, slips or discomfort during use. Some elements, such as window frames, have specific regulations that must be followed. In these cases, it is interesting to design with practicality in mind rather than the beauty of the equipment.The force required for the drive must be sufficient for an adult to be able to use it, the limits are 10 N for force and 20 N.m for torque.Taking into account that children will also use this equipment is important, so it is worth opting for items that also provide safety for them. Doors that give access to the outside environment can be made more difficult to open, such as a raised secondary latch, window frames can contain safety latches and the external frame of the window can already provide for the installation of safety nets, confined environments such as wardrobes and bathrooms can receive doors that do not close easily or that have fasteners to keep them open, and that contain latches with master keys, allowing them to be opened at any time if any of them get stuck.All information about the materials installed, how to use the equipment correctly and information about safety in the home should be described in the use and operation manual.

CHAPTER 8

Thermal performance

Buildings that are designed with thermal performance in mind not only use less energy to maintain the internal temperature, but also provide users with a better quality of life. Well-ventilated units with a mild temperature favour living and, above all, sleep for the resident, promoting greater satisfaction with their home. Thermal performance analysis can be done simply by choosing the position of the building and checking the thermal capacity of the façade elements, or in a more detailed way, using computer analyses with modelling of most of the natural phenomena that can affect the building.

Thermal performance depends on construction and utilisation characteristics. The former include the effects inherent to the building site, such as topography, orientation, local temperatures and humidity. The latter include the materials used, the number of floors and units per floor, the size of the rooms and window openings, the layout of the flats and the ceiling height. Even so, thermal satisfaction can only be guaranteed if the furniture is arranged correctly and depends on the number of people living there.

Brazil is divided into eight bioclimatic zones, for each of which there is a typical summer day and a typical winter day. These days contain set values for air temperature, relative humidity, wind speed and solar radiation. NBR 15575-1 Annex A2 shows a map of all the regions of Brazil and their respective bioclimatic zones. It also presents a table with the country's main cities and their data on daily maximum and minimum temperatures, daily temperature range, wet bulb temperature, solar radiation and cloudiness.

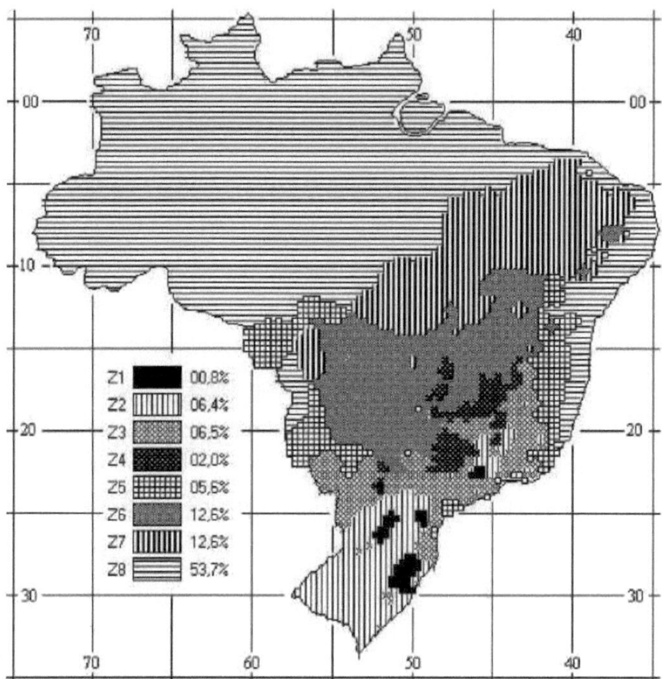

Figure 1[xi] : Brazilian bioclimatic zoning

Thermal assessment can be carried out in three ways: simplified, software simulation and on-site measurement:

- The simplified model is the result of meeting the building envelope criteria, based on the thermal transmittance and thermal capacity of the façade and roofs.
- The computer simulation should normally be carried out using the Energy Plus software[xii] . Developed by the US Department of Energy, it is a building simulation programme that allows integrated optimisation of energy and water systems in order to improve the building's thermal comfort. However, it is worth remembering that the standard does not deal with artificial conditioning, but rather natural conditioning, so it is necessary to use the appropriate tools in the programme to meet the requirements. This computer simulation is necessary when the analysis using the simplified model is not in line with the standard.
- On-site measurements are required by the standard in order to improve the results obtained in the previous models. They can be carried out on site with prototypes, or in existing buildings close to the site as long as they are similar to the building to be built.

The values referred to for transmittance and thermal capacity must comply with the limits established by the standard. In the case of walls, the materials used and the construction method are important for improving the wall's properties.

Solid concrete walls with a thickness of less than 10 cm do not meet any of the standard's requirements. Solid brick walls with or without cladding meet the specifications of zones 3 and 8 as long as they are painted in light colours. Ceramic brick walls with holes and cladding meet the standard in all regions. It is worth remembering that it is important to check the material's properties with the manufacturer and ensure compliance with the standard.

In the case of roofs, clay or fibre cement tiles without lining do not meet the requirements of the standard. Clay tiles, fibre cement with a wooden lining or a mixed slab can meet the minimum requirements. Roofs with clay or fibre cement tiles, polished aluminium foil and wooden lining meet the intermediate performance level. In order to guarantee the higher level, it is necessary to carry out the other analyses on the materials applied at the intermediate level.

In general, sandwich tiles made up of steel or aluminium elements filled with polyurethane meet the requirements of the standard at the top level. Companies such as Ethernit[xiii] and TermoRoof[xiv] offer products that are guaranteed to meet the standard and also have a finish on the underside, eliminating the cost of installing a ceiling tile over the roof. These products may cost more than conventional ones, but the compliance with the standard guaranteed by the manufacturer may be enough to avoid the need to carry out a computer simulation analysis and guarantee the building's superior performance.

In any case, when a selected product does not meet the requirements of the simplified analysis, it is possible to carry out the study by computer analysis and check that the choice meets the desired level of performance. The program is able to analyse the temperature variation from solar radiation, thermal conduction, heat generation from internal sources and convection, based on previously added material and environmental data. This makes it possible to produce a report with the thermal conditioning loads and the temperature of each room in the project, thus verifying compliance with the standard.

In cases where solar radiation is an important factor in thermal comfort, it is worthwhile for the designer to provide for the installation of curtains or blinds that prevent up to 50 per cent of this radiation from entering the room. When this cut is included in the computer analysis, it can make projects that were not previously allowed feasible. However, the use, operation and maintenance manual must specify what properties the curtains and blinds must have in order to guarantee this reduction in incoming solar radiation. In general, blackout curtains are able to cater for this reduction in sunlight and are cheaper for the resident.

Natural ventilation is one of the main factors in maintaining temperature, which is why the standard provides for a minimum opening that can vary from 7 to 12 per cent of the room's floor area according to the bioclimatic zone, and in certain places this area must be sealed off.

A study[xv] published in the construction forum of the Brazilian Institute for the Development of Architecture explained the minimum dimensions for each room using criteria based on technical standards. With these values, it is possible to establish the minimum size of a square opening for ventilation in each room, as shown in the table below.

Environment	Floor area [m]^2	Opening area [m]^2	Side [cm]
Kitchen	4	0,48	70
Service area	4	0,48	70
Bathroom	2,5	0,3	56
Washroom	1	0,12	36
Master bedroom	8	0,96	98
Secondary bedrooms	6	0,72	86
Wardrobe	4	0,48	70
Living room	8	0,96	98

Table 3: Relationship between floor areas and windows

It is up to the designer to choose and check window frames with sufficient openings to meet the minimum area required by the standard.

CHAPTER 9

Acoustic performance

One of the biggest causes of disagreements between residents of condominiums is noise, not because of parties held in the flats, but because of everyday noises. Wearing high heels, loud music or small renovations carried out by the resident themselves cause stress for the neighbours. With this in mind, as well as noise caused by cars in the garage or children playing in communal areas, a specific chapter was inserted in the performance standard to guarantee the acoustic comfort of the user.

As a wave, sound needs a medium to propagate through, and in solid media it propagates at higher speeds than in air. The acoustic pressure caused by the sound wave on the listener's eardrum is measured in Decibels (dB). The standard presents a comparative table of common pressures and situations audible to humans.

Pa	dB	Correspondence
20	120	Pneumatic hammer, aeroplane turbine
2	100	Vehicles with open exhausts
0,2	80	Avenue with heavy traffic
0,02	60	Radio at normal volume
0,0002	20	Limit for peaceful rest
0,00002	0	Hearing limits for young people

Table 4[xvi] : Comparative table between pressures (Pa) and decibels (dB)

The standard concentrates its criteria on sound and noise propagated by the impact of bodies on slabs and airborne propagation, seeking to improve the acoustics of floors and enclosures such as façades and roofs. Parameters are established for walls divided between two units, but not between walls in the same unit, and it is up to the designer whether or not to apply them.

9.1 Verification methods

The standard establishes various ways of checking the intensity of sound and the building's performance against its effects. For airborne sound, it is possible to check in the field using the engineering, simplified or precision method. Impact noise can be checked in the field.

9.1.1 Checking the airborne sound field using the engineering method

This test is standardised by ISO 140-5, which comprises an in situ measurement of the insulation against airborne sounds of façades and façade elements. It prescribes two measurement methods, the element method and the global method. The element method is used to assess the sound reduction of an individual façade element, such as a window or balcony door. The global method, on the other hand, is designed to assess the difference in exterior and interior sound levels in real traffic conditions, and can use the traffic itself as a sound source or loudspeakers that simulate a specific condition. This test results in a parameter that symbolises the weighted standardised level difference (DnT,w).

9.1.2   Checking the airborne sound field using the simplified method

When it is not possible to carry out the previous test, the simplified method can be used. It is more precise than the engineering method but can be used to determine the parameters of individual elements or the overall structure. Standardised by ISO10052, its results are presented in dB in a parameter defined as the standardised level difference weighted at 2 m (D2m,nT,w).

9.1.3   Laboratory test for airborne sounds using the precision   method

Unlike previous methods, this laboratory test studies each building element individually and calculates the relationship between them to determine overall performance. Standardised by ISO10140-2, the result is the weighted sound reduction index (Rw) measured in dB.

9.1.4.   Field verification of noise caused by impacts on floors

ISO 140-7 describes the procedure for analysing noise caused by walking or falling objects on the slab between floors and on the roof of buildings. It results in the weighted standardised impact sound pressure level (L'nT,w) expressed in dB.

9.2.   Performance criteria

Performance levels are improved as the sound insulation of structural elements and the overall ensemble improves. There are tables specifying limits for the items described above. The performance values measured in the field are generally lower than those obtained in the laboratory, the difference being due to the boundary conditions and the procedures adopted. The acoustic insulation of walls is mainly determined by their mass, but for perforated masonry it is also important to check the geometric aspect and the arrangement of the holes.

9.2.1.   Dnt,w - Weighted standardised level difference between environments

Element	Dnt,w (dB)	Level of performance
Wall between autonomous housing units (twinning) without bedroom	40 a 44	M
	45 a 49	I
	≥ 50	S
Wall between independent living units (twinning) with bedroom	45 a 49	M
	50 a 54	I
	≥ 55	S
Blind wall of dormitories between housing units and common areas of occasional transit	40 a 44	M
	45 a 49	I
	≥ 50	S

	30 a 34	M
Blind wall of living rooms and kitchens between a housing unit and common areas of occasional transit	35 a39	I
	≥ 40	**S**
	45 a 49	M
Blind wall between housing units and communal areas where people stay.	50 a 54	I
	≥ 55	**S**
	40 a 44	M
Set of walls and doors of separate units separated by the hallway	45 a 49	I
	≥ 50	**S**

Table 5[xvii] : Weighted standardised level difference between environments (Dnt,w)

9.2.2. Rw - Weighted sound reduction index of construction components of seals between rooms

Element	Rw (dB)	Performance level
	45 a 49	M
Wall between autonomous housing units (twinning) without bedroom	50 a 54	I
	≥ 55	**S**
	50 a 54	M
Wall between independent living units (twinning) with bedroom	55 a 59	I
	≥ 60	**S**
	45 a 49	M
Blind wall of dormitories between housing units and common areas of occasional transit	50 a 54	I
	≥ 55	**S**
	35 a39	M
Blind wall of living rooms and kitchens between a housing unit and common areas of occasional transit	40 a 44	I
	≥ 45	**S**
	50 a 54	M
Blind wall between housing units and communal areas where people stay.	55 a 59	I
	≥ 60	**S**
	45 a 49	M
Set of walls and doors of separate units separated by the hallway	50 a 54	I
	≥ 55	**S**

Table 6[xviii] : Weighted sound reduction index of inter-room sealing components (Rw)

9.2.2.1. Indicative weighted sound reduction index values for wall systems

The Technological Research Institute, in partnership with Unicamp and the University of Coimbra, has produced a study listing the types of wall and the Rw associated with each one. It helps designers determine the best element to build in each environment.

Wall type	Block width (cm)	Coating per side (cm)	Rw (dB)
Hollow concrete blocks	9	1,5	41
Hollow concrete blocks	H,5	1,5	42
Hollow concrete blocks	14	1,5	45
Hollow ceramic blocks	9	1,5	38
Hollow ceramic blocks	H,5	1,5	40
Hollow ceramic blocks	14	1,5	42
Solid fired clay bricks	11	2	45
Solid fired clay bricks	15	2	47
Solid fired clay bricks	11 + 11	2	52
Solid reinforced concrete wall	5	0	38
Solid reinforced concrete wall	10	0	45
Solid reinforced concrete wall	12	0	47
Drywall	2 sheets + glass wool	0	41
Drywall	4 plates	0	45
Drywall	4 sheets + glass wool	0	49

Table 7[xix] : weighted sound reduction index for wall systems

Based on this study, the following comparison can be made between the types of wall and the construction element and its performance index applied.

Drywall			Solid reinforced concrete wall			Solid fired clay bricks			Hollow ceramic blocks			Hollow concrete blocks			
4 ch. + wool	4 ch	2 ch. + wool	12	10	5	11 + 11	15	11	14	11,5	9	14	11,5	9	Element
M	M	NA	M	M	NA	I	M	M	NA	NA	NA	M	NA	NA	Wall between independent living units (twinning) without a bedroom
NA	NA	NA	NA	NA	NA	M	NA	NA	NA	NA	NA	NA	NA	NA	Wall between self-contained housing units (twinning) with bedroom
M	M	NA	M	M	NA	I	M	M	NA	NA	NA	M	NA	NA	Blind wall of dormitories between housing units and common areas of occasional transit
S	S	I	S	S	M	S	S	S	I	I	M	I	M	M	Blind wall of living rooms and kitchens between a housing unit and common areas of occasional transit

NA	NA	NA	NA	NA	NA	M	NA	NA	NA	NA	NA	NA	NA	NA	Blind wall between housing units and communal areas where people stay.
M	M	NA	M	M	NA	I	M	M	NA	NA	NA	M	NA	NA	Set of walls and doors of separate units separated by the hallway

Table 8: Comparison between wall types and construction element

As you can see, the only system that fulfils all the requirements would be a double wall of solid clay bricks, but for each location the wall can be designed according to the desired specification. If only the mortar cladding is used, the only possible performance will be average. In environments where this performance can be achieved with other, simpler wall types, it is recommended to use them in order to reduce costs. If it is necessary to increase performance, it is permissible to provide an acoustic seal to improve the installed walls. Remember that these same walls must also fulfil other criteria of the standard.

9.2.3 Weighted standardised level difference of the external seal for field tests (D2m,nT,w and Rw)

These tests are carried out when the room is closed and the influence of the external environment on the room's acoustic comfort is verified.

Noise class	Location	D2m,nT,w (dB)	Rw (dB)	Level of performance
I	Housing located far from sources of natural noise of any kind	≥ 20	≥ 25	M
		≥ 25	≥ 30	I
		≥ 30	≥ 35	S
II	Housing located in areas subject to noise situations that do not fall into classes I and III	≥ 25	≥ 30	M
		≥ 30	≥ 35	I
		≥ 35	≥ 40	S
III	Housing subject to intense noise from means of transport and other types of noise, in accordance with legislation	≥ 30	≥ 35	M
		≥ 35	≥ 40	I
		≥ 40	≥ 45	S

Table 9[xx] : Weighted standardised level difference of the external seal for field tests (D2m,nT,w and Rw)

As these tests refer to façades and roofs, it is very important to foresee the presence of gaps that can influence the loss of acoustic performance. Factors that influence the reduction in insulation are dry joints in the masonry, irregularities in the masonry laying material in the slabs and vi- gas, poorly sealed windows or windows with poorly installed glass and faulty grouting between walls and frames. To avoid these problems, the sealing work must be carried out with quality control and with the right materials for each job.

9.2.3.1 Indicative weighted sound reduction index values for doors and windows

The IPT has also drawn up tests for the main door and window systems. By adapting this table to the requirements of the standard, it is possible to assess which systems are best suited to the locations and their performance levels.

System	Rw (dB)	Noise class		
		I	II	III
Smooth door with hollow core, surface mass of 9 kg/m^2, untreated on the jambs	18	NA	NA	NA
Solid door, surface mass of 60 kg/m^2, with treatment on the jambs	28	NA	NA	NA
Aluminium sliding window, two leaves, 4mm glass	20	NA	NA	NA
Aluminium sliding window, one pane of 4 mm glass, two shutters	19	NA	NA	NA
Aluminium sliding window, integrated with curtains, two panes of 4 mm glass	26	M	NA	NA
Aluminium sliding window, two panes of glass 3 mm	23	NA	NA	NA
Aluminium sliding window, one pane of 3 mm glass, two shutters	16	NA	NA	NA
Maxim-air aluminium window, 4 mm glass	27	M	NA	NA
Maxim-air steel window, 4 mm glass	24	NA	NA	NA
Sliding steel window, one pane of 4 mm glass, two shutters	15	NA	NA	NA
Sliding steel window, four panes of 4 mm glass	16	NA	NA	NA
Aluminium opening window, 6 mm and 4 mm thick double glazing, 10 mm air gap between panes	30	I	M	NA
Aluminium opening window, 6 mm and 4 mm thick double glazing, 12 mm air gap between panes	36	S	I	M

Table 10[xxi] : weighted sound reduction index for doors and windows

As you can see, few common items meet the requirements of the standard. To solve this problem, it's worth looking for windows with glass thicker than 4 mm and improving the external treatment of solid core doors.

9.2.4 Airborne noise insulation (DnT,w) and impact insulation (L'nT,w) of accessible floors and roofs

Insulation between slabs is the main point of attention in the acoustic performance chapter, as it is largely responsible for user comfort.

Element	DnT,w (dB)	Level of performance
Floor system separating autonomous housing units from sleeping areas	45 a 49	M
	50 a 54	I
	≥ 55	S
Floor system separating self-contained housing units from communal areas of occasional transit	40 a 44	M
	45 a 50	I
	≥ 50	S
Floor system separating autonomous housing units from common areas where people stay	45 a 49	M
	50 a 54	I
	≥ 55	S

Table 11[xxii] : Airborne noise insulation (DnT,w)

Considering the slab without any cladding, the minimum thickness that meets the airborne noise requirements is 15 cm.

Element	L'nT,w (dB)	Performance level
Floor system separating autonomous housing units positioned on different floors	66 a 80	M
	56 a 65	I
	≤ 55	S
Accessible roofs or collective area floors over autonomous housing units	51 a55	M
	46 a 50	I
	≤ 45	S

Table 12[xxiii] : Impact noise insulation (L'nT,w)

Impact noise insulation is defined by the density of the material, unlike airborne noise insulation, which is defined by mass. Therefore, slabs without any cladding must have a thickness of 12 cm to meet these criteria, and there is no increase in insulation with greater thicknesses.

In both cases, compliance with top-level performance criteria is defined by the installation of acoustic protection blankets and good quality floor coverings. Blankets of 5 to 20 mm made of glass wool, stone wool or rubber are commonly used for this purpose. Aubicon[xxiv] has an easy-to-apply 5 to 8 mm blanket made from recycled rubber and adds little height to the floor, achieving a L'nT,w of 49 dB, Superior performance for slabs between units and Intermediate for roofs. PROMAFlex[xxv] has a blanket made of foam and polyethylene that meets the standard and is certified by the Institute of Research and Technology, the same institute that carried out the research for the parameters required by the standard.

9.3 Acoustic insulation against noise from sanitary equipment

The standard also mentions some criteria for insulation against sanitary equipment as a non-mandatory item. These noises are related to discharges from neighbouring units, the passage of water in plumb lines inside walls, sounds related to the operation of tanks. The maximum values of the equivalent continuous sound pressure level ($L_{Aeq,nT}$) must be measured in the field and inside the dormitories. The reference values are:

LAeq,nT (dB)	Performance
≤ 30	S
≤ 34	I
≤ 37	M

Table 13[xxvi] : maximum equivalent continuous sound pressure level values ($L_{Aeq,nT}$)

The way to promote adequate performance is partly design and partly construction. The design is related to the position of the rooms in each unit, avoiding the proximity of rooms to plumbing passage areas or near water tanks, when necessary, making the Shafts with the appropriate acoustic insulation. It is the building contractor's responsibility to opt for hydraulic equipment that is less susceptible to vibrations, such as plastic or flexible metal pipes, and when necessary to wrap larger pipes with acoustic lining. Heavy equipment such as central air conditioning units or water heaters can be installed on dampers to prevent vibration.

CHAPTER 10

Light performance

Natural lighting in homes is not only important for the homeowner's bank account, but also for their health. Since the sun is a source of vitamin D and is able to illuminate rooms for 12 hours every day, a chapter of the standard was designed to focus solely on the comfort and savings generated by the presence of natural light inside homes.

10.1. Required levels

Considering only natural lighting, the standard provides for the Lux levels available in each room of the home.

Environment	General illuminance for performance levels (Lux)		
	M	I	S
Living room Bedrooms Kitchen Service area	≥ 60	≥ 90	≥ 120
Bathroom Corridor and stairs Garages	Not required	≥30	≥45

Table 14[xxvii] : Required lighting levels

Light performance can be achieved in the following ways:

- Windows with large openings
- Orientation of units favouring sunshine
- Use light colours when painting walls and ceilings and choosing flooring
- Arrangement of mirrors inside communal areas or housing units
- Respect a certain distance between buildings or slopes, avoiding the early breaching of units

It's worth remembering that by favouring natural lighting, ventilation is also improved, making it easier to meet this requirement of the standard.

The standard also stipulates a minimum level of artificial light in each environment, which can preferably be provided by LED or fluorescent lamps, in order to reduce the costs of use and the maintenance required. Copel, the energy generation and distribution company from Paraná, carried out a study[xxviii] on the light bulbs available in Brazil and published the results in its periodical Espaço Energia.

LED	Fluorescent Compact	Incandescent	**Model**

5,81	11,97	40,70	**Power (W)**
35.000	35.000	35.000	**Hours Used**
97,59	201,06	683,65	**Energy Cost (R$)**
35.000	8.000	750	**Service life (Hours)**
1	5	47	**Lamps used**
79,5	12,90	2,40	**Unit Cost (R$)**
79,50	64,50 2	112,80	**Replacement Cost (R$)**
177,09	265,56	796,45	**Total Cost (R$)**

Table 15[xxix] : Lamp comparison

Having seen this study, it is clear that the difference between the types of light bulb is that choosing the most economical one is also part of a better-performing building. Guidance on the type of lighting, its maintenance and useful life should be available in the owner's manual.

CHAPTER 11

Watertightness

The presence of water inside a home can be one of the main factors in reducing its useful life. Moisture can cause mould and fungus to grow, leading to respiratory illnesses in residents. In addition, building elements in the presence of water can suffer from corrosion or leaching, reducing their durability. With this in mind, the standard established various parameters for waterproofing housing units.

In order to better classify the criteria, it is important to understand the difference between wet areas and wettable areas. The former is characterised by the presence of a sheet of water in its use, such as shower cubicles, service areas and roofs. The latter refers to areas that can be wet, without the formation of a sheet, such as bathrooms, kitchens and balconies.

Watertightness must be guaranteed to prevent rainwater from seeping through roofs, moisture from the ground and possible leaks between neighbouring walls, external walls and internal walls between "dry" and "wet" rooms.

11.1. Floor systems

Rising water coming from the capillarity of the soil or from the levelling of the water table cannot enter the house, so it is necessary to take protective measures. In general, damp problems occur in clay soils, and can be protected by waterproofing the subsoil and applying specific blankets to the base of walls and pillars that are in direct contact with the ground.

11.1.1. Wet areas

As these areas will be in constant contact with water, it is recommended to carry out a waterproofing and drainage process. The standard requires that in

up to 72 hours for the surface to dry out when in contact with a 10 mm sheet, thus preventing the appearance of damp.

When the wet area has no waterproofing, ceramic tiles can be made watertight as long as the grouting is waterproof and there are no flaws in the installation.

To ensure watertightness in non-waterproofed areas, it is ideal to make the floors with a 1% slope for indoors, 0.5% for outdoors and 2% for shower cubicles. This slope, when directed towards a drainage system, prevents puddling, which is responsible for the appearance of damp.

11.1.2. Wet areas

As they are not considered watertight, all wet areas must be described in the use, operation and maintenance manual. This manual should contain the best way to sanitise these areas to avoid water accumulation. It is also worth providing drains with some waterproofing around them in these areas, to guarantee watertightness even if the guidelines described are not followed.

11.2. Facades and internal walls

The water that affects the façades comes not only from the local rainfall but also from the speed and direction of the winds in the area. The performance of façade elements is measured through tests in which these elements are subjected to the presence of water and wind for a set period of time, the appearance of stains is quantified and performance is then measured. For a 7-hour test, if up to 5 per cent of the tested area is stained, the performance is Average, while if there are no stains, the performance can be Intermediate or Superior. Therefore, the choice of façade cladding material is extremely important in guaranteeing performance.

Water penetrates façades through cracks, detachment of façade slabs from the wall, poorly executed grouting and splashing on sills. This requires the correct application of materials and quality laying of the ceramic tiles or stones that make up the façade, avoiding cracks, movement or loss of the adhesive effect with the wall. In very high buildings, blind gables often receive real cascades of water, which can cause the cladding to leach. To avoid this effect, it is worth designing small recesses or small elevations in the wall to break up the energy of the water.

The correct maintenance of façade elements and especially the cleaning methods must be written in the use, maintenance and operation manual, thus guaranteeing performance.

For internal walls, the same recommendations apply as for façades, but in wet areas, it is worth providing waterproof areas, especially near sinks, as these are places where small amounts of water splash up.

11.3. Covers

The watertightness of roofs depends on the materials used to conduct the water, their slope, the way they are installed and the capacity of the drainage system. This capacity must be calculated according to the local rainfall index, in accordance with NBR 10844. The entire top slab must be waterproofed with mandas or membranes accepted by NBR 9574 and NBR 9575.

Roof tiles must be watertight, free of drips and stains, and can be made of any material, as long as it meets the requirements of the standard. In the case of tiles made of porous material (clay), the application of paints, resins or other forms of waterproofing is permitted.

Check for barriletes, shafts, ventilation pipe ends, chimneys, air intakes and other possible openings that could allow water to enter the building and carry out waterproofing or ways to prevent water from entering.

The use, maintenance and operation manual must provide the necessary information on the replacement of roofing elements, such as tile fasteners, waterproofing blankets and their respective maintenance and periodic cleaning. The cleanliness of gutters and the rainwater collection system should be a determining factor when choosing a system.

11.4 Hydrosanitary system

All the components of the water and sanitation system must be watertight: water tanks, reservoirs, pipework, equipment, manholes, delay boxes, collection boxes, grease traps and so on. For water tanks and reservoirs, the correct sizing and execution of both the structure and the waterproofing are critical to prevent any cracks and allow water to leak out. Pipes must also be installed with care and allow access for maintenance whenever possible, as leaks will occur inside the structural elements of the building and are therefore more difficult to identify and correct.

CHAPTER 12

Durability

Sustainability, the hot topic at the moment, can also be applied to the development of construction. For a long time, projects were designed and built with the aim of meeting deadlines and reducing costs, as well as looking at the beauty of the development and its attractiveness for sales. However, with the concept of sustainability on the rise, building durability has become one of the most important points in the construction process. Durability is defined by the standard as the useful life of the development. Useful life is understood as the time during which the building maintains its performance conditions unchanged, with periodic maintenance.

In order to think about durability and useful life, it is worth considering the change in strategy when considering the cost of construction. Initially, the cost was related to the development of projects and the construction itself. Nowadays, the cost also includes the maintenance required by the building throughout its useful life and any repairs that may be necessary.

Meeting the durability requirements of the standard involves three lines of thought. The first relates to the impacts caused by a possible system failure, the second to ease of maintenance (maintainability), and the third to the cost of correcting the damage and its influences. This durability is quantified by the Project Useful Life (PUL), which can be longer or shorter depending on the level of performance desired, but only the minimum is mandatory. Meeting higher standards is of interest to developers because it adds value to the development, but it is mainly of interest to the construction industry because it drives new, more economical construction techniques, makes it possible to reduce costs during maintenance and corrections, and creates competitiveness in the market with better quality products.

As VUP is a theoretical parameter, in order to guarantee that the useful life (VP) is achieved in reality, it is necessary for everyone in the construction chain to prioritise quality when carrying out their work. Designers are responsible for creating quality, well-detailed projects that avoid errors during construction. It is the builders' duty to carry out quality services and monitor the correct use of the materials employed. Suppliers are responsible for developing suitable materials and guaranteeing their quality throughout the life of the project. Owners are responsible for looking after their property, carrying out periodic maintenance and inspections whenever possible, identifying possible faults at an early stage. All that's left is for developers to be willing to pay a little more to ensure the quality of their developments.

12.1. Design life

To help project developers, the standard suggests a mandatory minimum value for each building component and higher values that add value to the development. An indicative table has been drawn up showing each component and the useful design life for each performance level.

System	VUP [years]		
	M	I	S
Structure	≥ 50	≥ 63	≥ 75
Internal floors	≥ 13	≥ 17	≥ 20
External vertical sealing	≥ 40	≥ 50	≥ 60
Internal vertical seal	≥ 20	≥ 25	≥ 30
Cover	≥ 20	≥ 25	≥ 30
Plumbing	≥ 20	≥ 25	≥ 30

Table 16[xxx] : Design lifetimes

Whenever the VUPs are not specified in the project, the deadlines for minimum performance will be taken into account.

The useful life begins as soon as the work is legally completed, i.e. after approval of the "habite-se" or similar documents. It can be said that the useful life has been reached when 50 per cent of its time has elapsed without the need for major impact interventions. These major interventions are classified in table C.3 in NBR 15575-1, but basically represent interventions in which the cost of maintenance was higher than the cost of construction. It is valid

also check the VUP for compliance with existing technical standards, and responsibility can be transferred to a third party, such as an insurer or the building manager, for example.

12.2 Durability of the building and its parts

Durability compatible with the VUP can be achieved in practice by using materials and components that comply with the standard, by carrying out the work with strict quality control and good practices, and by correctly using the information described in the use and operation manual, guaranteeing efficient maintenance.

Ensuring durability also involves anticipating and avoiding pathologies, and if they do occur, identifying them quickly and treating them before the damage becomes more damaging. Some special precautions can be taken to avoid such incidents:

Reinforced concrete structures require a lot of human labour to build, which is why failures are more frequent. During execution, it is worth observing the correct cover, guaranteeing the quality of the concrete, especially checking its fck, using the right reinforcement, taking care when curing to ensure the necessary humidity and avoid exposure to high temperatures, adequate vibration during concreting to avoid disintegration of the components, removal of the forms and props at the right time, among others.

Steel structures suffer greatly during transport and from oxidation after installation, so care must be taken to ensure that the parts are properly protected during the journey between the factory and the construction site, as well as correct storage in the field. It is recommended to use welds and screws with anti-corrosive treatments as these are fragile points in the structure. The assembly of the parts is crucial to the overall alignment of the structure and its strength, and must be carried out by trained staff and checked meticulously.

It is worth remembering that steel is an excellent electromagnetic conductor, which is why lightning protection systems must be added to the project.

The use of timber structures is not very common in the country, but it is important to emphasise the importance of using suitable and legalised specimens. Connections composed of other materials must be well studied

avoiding waterlogging and warping of the wood. The pieces chosen should preferably be clean of knots and other fragile points. As it is an organic material, it is important to prevent the action of bacteria and animals that could damage the structure, as well as the effects caused by humidity and its variation with paints and varnishes suitable for each purpose.

Floor systems and vertical fences suffer a lot from water run-off, so it's a good idea to check that the rooms are levelled and that the drainage systems are correctly installed. Whenever possible, some form of waterproofing can be carried out, especially when housing units share the same slab, such as floors and ceilings. The materials used, such as ceramics or paints, undergo visual changes when they come into contact with water and solar radiation, so it's worth making sure that these elements are proven to comply with the standard.

Window and door frames are directly used by the user. As they are mobile structures, their component material must withstand minor user negligence such as unnecessary forces when opening and closing doors and windows, or alternative ways of keeping the element in a suitable position. Watertightness is important, it is recommended that the application of water seals is clearly explained in the use, operation and maintenance manual. Whenever possible, specifying drip rails, sills and roofs ensure better performance of the element.

Leaks in inaccessible places are complicated to identify and resolve, so it's important to avoid using different metals due to the formation of piles that can start a corrosive process, to size and design the paths correctly to avoid water hammer or insufficient flow, to design and build water tanks and reservoirs to avoid the appearance of cracks, among other things.

The standard also emphasises the great importance of carrying out periodic inspections and maintenance on the elements of the dwelling in accordance with the instructions in the use, operation and maintenance manual. This activity guarantees the durability of the materials used and increases the possibility of achieving the design service life.

CHAPTER 13

Maintainability

As one of the main factors in housing performance, regular maintenance carried out under the maintainability conditions specified in the project creates a closer relationship between the user and their home. The user's thinking that their home is unattainable and serves to protect them must be reformulated: their home is one of their most precious possessions and needs to be looked after like any other asset, and only then will it be able to safely fulfil the purposes for which it was built. If the user is a layperson in the field of design and construction, it is important that there is a way that is easy for the owner to understand and apply over the years. The document called the use, operation and maintenance manual is, in fact, the instruction manual for the home, it talks about its correct use, how it was built and how to take good care of it.

13.1. Use, operation and maintenance manual

Considered to be the user's main tool for guaranteeing the useful life of the building, the use, operation and maintenance manual is a document generated by designers and builders that includes all the information pertinent to the building, its components and maintenance and repair methods. It is important that this document describes:

Warranty certificates for materials, furniture and equipment, information on technical assistance for all of them, as well as a guarantee of compliance with the standard and the VUP. This information can be sourced from suppliers or obtained by the designer or builder from tests pre-defined in the standard.

The descriptive memorial, especially for installation projects, containing the specifications of each material and the place of application. If possible, the presence of as-built projects is of great value, especially if the project can be expanded in the future.

A list of suppliers of materials, designers who have worked on the project and utilities that have participated in the construction or design. Up-to-date contact details for all these participants will help the building manager when they need relevant information about the project.

Information on the utilisation and use of the property in all its systems such as: plumbing, electrical and electronic, SPDA, mechanical ventilation and heating, automation, communication, fire, foundations, structure, seals, coverings, floors, roofs, gardens and leisure areas, window frames and glass, interconnections with the public network. This information is obtained from supplier guidelines and material specifications. They can be written as clearly as possible, where necessary with a glossary of words that are less familiar to laypeople.

Develop a maintenance programme for equipment and installations, especially at fragile points that could cause damage if forgotten. It may be interesting to draw up standard checklists for the systems and show building management how to use them.

Information about preserving the environment, such as encouraging selective collection, suggesting possible ways to save water and energy both in the housing units and in common areas, and other tips that help the owner's administration.

Caring for the security of the building, with regard to employee training, correct use of security systems, intercom systems and internal camera circuits.

Presentation of the need to train employees to operate more complicated machinery or systems, preferably with contacts of qualified professionals for this training. Suppliers usually provide this type of service themselves.

The building's legal documentation is also important, information on approval and legalisation processes, as well as registrations with notaries, ARTs for projects, land deeds and other documents of legal value.

CHAPTER 14

Overview of low-cost construction

The history of housing in Brazil is directly linked to its history. Since its discovery, the country has been treated as a second option, creating a lasting culture that our country is inferior to any other. This culture caused the country's administrators to be somewhat careless, which during its expansion led to uncontrolled growth, both in terms of infrastructure and population, creating the emergency society we live in today.

In the early years of our country's life, it was treated only as a source of raw materials for the metropolis, being occupied precariously by labourers, adventurers, renegades from Europe and those who owned small towns in their hereditary captaincies. It was only at the beginning of the 16th century, with the discovery of sugar cane, that the first coastal cities began to emerge: Salvador (the capital of the colony), São Sebastião do Rio de Janeiro, Filipéia de Nossa Senhora das Neves (João Pessoa) and Sergipe d'El-Rey, all of which arose because they were the end points of the sugar cane export routes and transport ports for the metropolis.

At the end of the 17th century, the decline of sugar and the discovery of gold led to a large migration route to the region currently represented by the state of Minas Gerais. This gold rush promoted a large migration of Portuguese settlers and those from other Portuguese colonies, starting the country's expansion into its interior and considerably increasing the number of port cities. The urban growth of these cities took place mainly around the ports and commercial centres, as they were the points through which the food produced in the interior arrived, and where opportunities for enrichment arose, the dwellings were precarious, very close to each other and had the minimum comfort necessary for housing.

At the beginning of the 18th century, with the Napoleonic invasions of Europe, Brazil became a united kingdom of Portugal, mobilising the migration of wealthy settlers and thus starting the first urban planning process in the country. With the presence of the "wealthy" came palaces, gardens, theatres, water supply systems, water and sewage systems, etc.
It also led to improvements in the health and culture of the main cities. This need for urbanisation led to the creation of jobs in the capital, now Rio de Janeiro. Previously uninhabited areas were developed, causing major urban sprawl in the city and a high level of migration to the south-east of the country.

The development of coffee-growing in the Paraiba valley led to a boom in Rio de Janeiro and São Paulo, their satellite cities and their trade centres. Even so, at that time only 6% of the country's population lived in the cities.

The first urban housing policy took place in the 19th century, with Hygienism. Rio de Janeiro, as the main city of the republic, concentrated a large part of the urban population in corners peripheral to the centre, known as favelas. Hygienism was a political decision that attempted to improve the health and sanitation qualities of these occupations, and it was not very well received.

The increase and inversion of the urban/rural population ratio occurred with the arrival of industries throughout the 20th century. The big cities received a large part of the rural population that could no longer work in the fields. This disorderly and sudden growth increased the rate of irregular housing of very poor quality and in risky situations, aggravating the problem of favelisation. From 1945 onwards, the Vargas government allowed the consolidation of various real estate portfolios and created aid and financing programmes led by the Social Service Institute of Brazil and the National Housing Bank. In 1956, the "Slums Law" was passed, which obliged urban centres to remove the population from the slums and create free low-income housing, and the Popular House Foundation came to assist in this process without much success. In the populist post-Vargas government, the transformation of the People's House Foundation into the Brazilian Housing Institute, with resources from the Trust Fund for Social Progress, the first housing units were financed with instalments equivalent to the minimum wage, and the low-income population was able to buy their own homes on their own.

It was only at the end of the 20th century and the beginning of the 21st century that there was a growing concern to provide quality housing for the less favoured, as the housing deficit was hampering urban growth. After the end of the military period and the re-democratisation of the country, Caixa Econômica Federal, which administers the Guarantee Fund for Length of Service, made it possible to buy better housing for the less well-off. This freedom boosted the construction of large-scale housing estates throughout the country, especially in the major metropolises.

The country's economic crisis hindered this growth for a while, but from 1995 onwards, under Fernando Henrique, the CEF began to release its financing again, new projects began to emerge, but there was still a lack of consolidation of rules for these types of projects and their financing on a national level. It was only in 2009, with the creation of the Minha Casa Minha Vida (My House, My Life) programme, that all residential financing projects were standardised, promoting a major expansion in this branch of construction. New specialised companies emerged from existing construction and development companies, taking their high-standard techniques and applying them to low-cost projects. Quality was improved, prices reduced, payment terms made easier, but the priority was always profit, which was acquired through large-scale production and thus compromised performance.

CHAPTER 15

Conclusion

The performance standard calls for building projects to be drawn up taking into account urban, geomorphological, environmental, climatic and usability aspects. Influenced by local resources, construction deadlines and the useful life of the project, materials are specified, as well as technologies and construction methods used in the building. Designs and execution must be drawn up at all times taking into account the possibility of faults and their correction, as well as periodic maintenance by the user. For this reason, the standard brings about a change in thinking throughout the construction chain: the idea that each participant's responsibility ends when their work is completed can no longer exist; everyone is responsible for the performance of the dwelling throughout its useful life, and they must ensure that the other parties feel responsible and act as "owners" of the project.

Civil construction in Brazil has undergone constant progress, and concern for performance has been one of the biggest advances so far. As well as enabling the development of new technologies, the time when the standard was launched directly affected those most in need of housing. Buyers who spend all their savings on their dream home expect to buy a product for life. Until then, this "life time" had not been quantified and many aspects of housing quality were forgotten or ignored in order to make more money. With the consolidation of the performance standard, there will be greater security when buying a home, the certainty that the product is of quality and will last as long as necessary to satisfy the owner's desires.

Adapting to the standard and building high-performance developments requires greater investment on the part of designers and builders. Designers need to better prepare their studies and drawings, with enough detail to avoid any doubts during execution. Construction companies need to seek the necessary savings in other aspects, without jeopardising quality and safety.

It is common knowledge that the Brazilian construction process is artisanal, generating a lot of material loss and waste. There is an opportunity for innovation and it is important to utilise it now. Project management is one of the greatest strategies for reducing construction costs. With it you can avoid compatibility problems, carry out physical and financial planning that is consistent with the reality of the project, taking into account the availability of labour and materials, as well as the logistical challenges of the work. Drawing up risk analyses at the start of project development and using them to mitigate possible unforeseen events on site helps to deal with problems that may arise, avoiding unnecessary costs. Generally speaking, designing with attention to detail and carrying out the work with rigid organisation and control generate great savings for the project. These savings can, and should, be used to improve construction quality.

It is high time to establish a major evolution in Brazilian construction, we need to mature the concepts of industrial production and extinguish the concept of the "Brazilian way". Guaranteeing the quality and performance of a development is an integration between design, construction and use, with everyone talking to each other from the start of the project, playing their part in the whole and guaranteeing the quality of their work. If one of the parts fails, the whole loses. For this reason, the new era in civil engineering is one of teamwork and technological development, seeking to improve together.

Bibliography

1. Brazilian Chamber of the Construction Industry. Performance of residential buildings: guidelines for compliance with ABNT standard ABR 15575/2013. Brasília, DF : Gadiolo Cipolla Comunicação, 2013. 624.07.

2. Brazilian Association of Technical Standards. Residential buildings - Performance - Part 1: General requirements. ABNT NBR 15575-1. s.l. : ABNT, 2013. Vol. 1. 91.040.01.

3. -. Residential buildings - Performance - Part 2: Requirements for structural systems. ABNT NBR 15575-2. s.l. : ABNT, 2013. Vol. 2. 91.040.01.

4. -. Residential buildings - Performance - Part 3: Requirements for flooring systems. ABNT NBR 15575-3. s.l. : ABNT, 2013. Vol. 3.

5. -. Residential buildings - Performance - Part 4: Requirements for internal and external sealing systems. ABNT NBR 15575-4. s.l. : ABNT, 2013. Vol. 4.

6. -. Residential buildings - Performance - Part 5: Requirements for roofing systems. ABNT NBR 15575-5. s.l. : ABNT, 2013. Vol. 5.

7. -. Residential buildings - Performance - Part 6: Requirements for plumbing systems. ABNT NBR 15575-6. s.l. : ABNT, 2013. Vol. 6.

8.   Cottrel, Michelle. Guide to the LEED Green Associate Exam. New Jersey : John Wiley & Sons, Inc, 2010.

9. Pontifical Catholic University of Rio de Janeiro. Norms for the presentation of theses and dissertations. Rio de Janeiro : s.n., 2001. 378.2420218.

[i] Brazilian Electrical and Electronic Industry Association. Comparison of heating systems. <http://www.abinee.org.br/>. Available at: < http://www. banhoeconom ico.com. br/banho2.htm >. Accessed on: 08/10/2013.

[ii] Sabesp; Neves, Maria Angélica Carneiro da Rocha, administrative manager of Hirosaki dedetização; De Castro, Sérgio Meira, director of condominiums at Secovi-SP. Sindiconet. Available at: < http://www.sindico- net.com.br/608/Informese/Limpeza-de-caixas-dagua >. Accessed on: 08/10/2013.

[iii] Resolution No. 307, of 5 July 2002< http://www.mma.gov.br/port/conama/legiabre.cfm?codlegi=307> Accessed on: 08/10/2013

[iv] Resolution   No.   448   of   18   January   2012 <http://www.mma.gov.br/port/conama/legiabre.cfm?codlegi=672>

ᵛ 1 Cottrel, Michelle. Guide to the LEED Green Associate Exam. Hoboken, New Jersey : John Wiley & Sons, Inc, 2010.

ᵛⁱ Sabesp; Monthly Report 3 Research Project Escola Politécnica / US- PxSABESP - June/96 and technical information from ASFAMAS. Available at: < Monthly Report 3 Research Project Escola Politécnica / USPx- SABESP - June/96 and technical information from ASFAMAS <http://site.sa- besp.com.br/site/interna/Default.aspx?secaold=145>. Accessed on: 21/10/2013.

ᵛⁱⁱ Unifrax, Fireproof Sealing Materials <http://www.uni- frax.com.br/produtos-materiais-selagem-fogo.asp> Accessed: 24/10/2013

ᵛⁱⁱⁱ Decflex, Ventilation equipment, fire dampers <http://www.decflex.com/engine.php?cat=76> Accessed: 24/10/2013

ⁱˣ Anti-Slip System, Anti-slip treatment < http://www.antis- lip.com.br/site/antislip/ >, Accessed on: 24/10/2013

ˣ Senate Bill No. 650 of 2011, Senator HUMBERTO COSTA <http://www.senado.gov.br/atividade/materia/getPDF.asp?t=98469&tp=1 >Accessed on: 05/11/2013

ˣⁱ ABNT NBR 15220 - Part 3

ˣⁱⁱ Office of Energy Efficiency & Renewable Energy, EnergyPlus Energy Sim- ulation Software <http://apps1.eere.energy.gov/buildings/energyplus/> Accessed 12/11/2013

ˣⁱⁱⁱ Ethernit, Sandwich Tiles <http://www.eternit.com.br/produtos/cober-turas/metalicas/telha-sanduiche> Accessed 17/11/2013

ˣⁱᵛ TermoRoof, Thermal line products <http://www.termoroof.com.br/pro- ducts> Accessed 17/11/2013

ˣᵛ Iberê M. Campos, Projeto de residência: um guia com medidas e áreas mínimas <http://www.forumdaconstrucao.com.br/con- teudo.php?a=6&Cod=112> Accessed 17/11/2013

ˣᵛⁱ Table E.2. - Annex E of standard NBR15575 - Part 1, page 65

ˣᵛⁱⁱ Table F.10, page 57 of NBR 15575-4

ˣᵛⁱⁱⁱ Table F.12, page 59 of NBR 15575-4

ˣⁱˣ IPT, Unicamp, SOBRAC, University of Coimbra

ˣˣ Table F.10 and F.12, page 57 and 59 of NBR 15575-4

ˣˣⁱ IPT, AFEAL, University of Coimbra

ˣˣⁱⁱ Table E.2, page 44 of NBR 15575-3

ˣˣⁱⁱⁱ Table E.1, page 43 of NBR 15575-3

xxiv Aubicon, Sound Soft Acoustic Blankets <http://www.aubicon.com.br/pro- dutos/mantas-acusticas-sound-soft. html> Accessed 20/11/2013

xxv PROMAFlex, PROMALaje Acoustic Insulation Blanket < http://www.promaflex.com.br/manta-de-isolamento-acustico-promalaje> Accessed on 20/11/2013

xxvi Annex B - Table B.2, page 34 of NBR 15575-6

xxvii Annex E - Table E.3, page 66 of NBR15575-1

xxviii Federal University of Paraná - Department of Civil Construction, Lactec - Institute of Technology for Development, Comparative study of light bulbs: incandescent, compact fluorescent and LED, ENERGY SPACE, ISSUE 18, APRIL 2013.

xxix Table 9 Simplified comparative cost analysis of the three types of bulbs, Energy Space Journal, Issue 8, April 2013

xxx Annex C, Table C.5, page 56 of NBR 15575-1

Buy your books fast and straightforward online - at one of world's fastest growing online book stores! Environmentally sound due to Print-on-Demand technologies.

Buy your books online at
www.morebooks.shop

Kaufen Sie Ihre Bücher schnell und unkompliziert online – auf einer der am schnellsten wachsenden Buchhandelsplattformen weltweit! Dank Print-On-Demand umwelt- und ressourcenschonend produziert.

Bücher schneller online kaufen
www.morebooks.shop

Printed by Books on Demand GmbH, Norderstedt / Germany